THE HISTORY OF AGRICULTURE

FROM SYMBOLS TO SCRIPTS

by Arthur Pendragon

Copyright © 2024 by Arthur Pendragon

All rights reserved. No part of this publication may be reproduced, distributed, or transmitted in any form or by any means, including photocopying, recording, or other electronic or mechanical methods, without the prior written permission of the publisher, except in the case of brief quotations embodied in critical reviews and certain other noncommercial uses permitted by copyright law.

TABLE OF CONTENTS

Introduction... 2

Chapter 1: The Dawn of Farming.. 5

Chapter 2: The Role of Animals in Early Agriculture................ 9

Chapter 3: Technological Innovations in Early Agriculture... 23

Chapter 4: Agricultural Tools and Techniques........................ 32

Chapter 5: Global Exchange of Crops and Animals................ 41

Chapter 6: The Industrial Revolution and Agriculture............ 47

Chapter 7: Scientific Advances in Agriculture......................... 62

Chapter 8: Modern Agriculture... 77

Conclusion... 87

Glossary of terms... 91

Timeline of key events... 94

References and further reading.. 96

INTRODUCTION

Agriculture has been a cornerstone of human civilization, shaping societies, economies, and the very landscape of the earth. From the dawn of farming to the advent of genetically modified crops, the story of agriculture is essentially the story of human progress. This book explores the evolution of agriculture, tracing its development from the earliest days of planting seeds and domesticating animals to the sophisticated farming techniques of the modern world.

Overview of the Importance of Agriculture in Human History:

The transition from a nomadic lifestyle of hunting and gathering to settled farming communities marked a profound change in human history. Agriculture allowed for the establishment of stable communities, the growth of populations, and the development of complex societies. It was the foundation upon which cities and civilizations were built. As people learned to cultivate the land and rear animals, they gained a measure of control over their food supply, reducing the uncertainties of hunting and gathering. This control over food production led to surplus harvests, which in turn enabled the growth of trade, the specialization of labor, and the rise of social hierarchies.

Agriculture not only provided sustenance but also spurred technological and cultural innovations. Irrigation systems, plows, and crop rotation techniques improved farming efficiency and productivity. Agricultural surplus supported artisans, traders, and scholars, fostering advancements in arts, sciences, and governance. The agricultural revolution was a driving force behind the development of writing, as record-keeping became essential for managing surplus and trade.

Contrast Between Hunting/Gathering and Farming:

Before the advent of agriculture, humans relied on hunting animals and gathering wild plants for food. This way of life was inherently unpredictable, dependent on the availability of game and the seasonal abundance of edible plants. Hunting and gathering required large areas of land and constant movement to follow food sources, leading to small, mobile groups of people.

In contrast, farming allowed for the cultivation of specific plants and the domestication of animals, providing a more reliable and controllable food supply. Settled farming communities could store surplus food, reducing the risk of famine and allowing populations to grow. Farming also enabled the development of permanent dwellings, leading to the establishment of villages and towns. This sedentary lifestyle facilitated the accumulation of goods, the construction of infrastructure, and the development of culture and governance.

The stability and predictability of farming compared to hunting and gathering cannot be overstated. Farmers could plan for the future, sowing seeds and raising animals with the expectation of harvests and offspring. This predictability allowed for long-term planning, investment in land and tools, and the growth of wealth and trade.

As Samuel Johnson, the English author, aptly stated, "Agriculture not only gives riches to a nation, but the only riches she can call her own." This quote encapsulates the fundamental importance of agriculture in creating and sustaining wealth. Unlike resources that can be exhausted or trade that can be disrupted, agriculture provides a renewable and reliable source of wealth. It underpins the economy of nations, supports the livelihoods of millions, and ensures food security for populations.

Agriculture is more than just the cultivation of crops and the raising of animals; it is the backbone of human civilization, providing the foundation upon which societies are built and thrive. This book will delve into the fascinating history of agriculture, exploring how it has transformed human life from the earliest times to the present day.

CHAPTER 1: THE DAWN OF FARMING

c.11,000–9000 BCE: First Farmers

The end of the last Ice Age around 11,000 BCE marked the beginning of a new era in human history: the dawn of farming. As the planet warmed and the glaciers retreated, vast areas of land in the Near East became fertile and habitable, setting the stage for one of the most significant transitions in human existence.

Development of Farming in Syria and Iran:

The regions of modern-day Syria and Iran were among the first to witness the birth of agriculture. These areas, part of a larger region known as the Fertile Crescent, offered an ideal environment for the development of farming. The Fertile Crescent, stretching from the eastern Mediterranean coast through the valleys of the Tigris and Euphrates rivers to the Persian Gulf, was a lush and verdant area with rich soils and abundant water sources.

In this fertile region, early humans began to experiment with the cultivation of wild grasses and the domestication of animals. This shift from a nomadic lifestyle of hunting and gathering to a more settled way of life was driven by several factors, including the availability of arable land, the abundance of wild cereals, and the changing climate that made these activities viable.

Archaeological evidence suggests that by around 11,000 BCE, small groups of people in Syria and Iran were already cultivating wild barley and wheat. These early farmers used simple tools made of stone and bone to till the soil and sow seeds. The transition to farming allowed them to produce a more reliable and abundant food supply, reducing their dependence on the unpredictable availability of wild game and plants.

Growth of Wheat and Barley in the Fertile Crescent:

The cultivation of wheat and barley in the Fertile Crescent was a pivotal development in the history of agriculture. These cereal crops were well-suited to the region's climate and soils, and they provided a stable and nutritious food source. Wheat and barley are hardy plants that can grow in a variety of conditions, making them ideal for early agricultural societies.

The domestication of these grains involved selecting and planting the seeds of the best-performing plants, a process that gradually led to the development of more productive and resilient crop varieties. This early form of selective breeding was crucial in increasing agricultural yields and ensuring food security for growing populations.

The ability to grow and store surplus grain allowed communities to settle in one place, leading to the development of permanent villages and the growth of population centers. These early farming communities were able to support larger populations than hunter-gatherer groups, and they laid the foundation for the development of complex societies.

In addition to providing a reliable food source, the cultivation of wheat and barley had other significant impacts. The surplus grain could be stored for future use, traded with neighboring communities, or used to support specialized labor and emerging social hierarchies. The ability to produce and manage surplus food was a key factor in the rise of civilization.

The dawn of farming in the Fertile Crescent was not an isolated event but part of a broader pattern of agricultural development that eventually spread to other parts of the world. The techniques and knowledge developed by these early farmers were passed down through generations, influencing agricultural practices in other regions and shaping the course of human history.

In summary, the period from around 11,000 to 9000 BCE was a transformative time in human history. The development of farming in Syria and Iran, and the cultivation of wheat and barley in the Fertile Crescent, marked the beginning of agriculture as we know it. This shift laid the groundwork for the growth of civilizations, the advancement of technology, and the complex societies that would follow.

CHAPTER 2: THE ROLE OF ANIMALS IN EARLY AGRICULTURE

c.10,000 BCE: A Farmer's Best Friend

Domestication of Dogs from Wolves:

Around 10,000 BCE, a significant milestone in the relationship between humans and animals was achieved with the domestication of dogs from wolves. This development was not only a testament to human ingenuity and adaptability but also a crucial step in the evolution of agricultural societies.

The Early Relationship Between Humans and Wolves:

The origins of the domesticated dog trace back to wolves, which were widespread across Eurasia. Early humans and wolves often inhabited the same territories and faced similar challenges, such as finding food and protecting themselves from larger predators. Over time, a symbiotic relationship began to form between humans and wolves.

Wolves are social animals with complex pack structures and behaviors that emphasize cooperation and communication. Early humans likely observed these traits and recognized the potential benefits of forming a closer bond with wolves. Mutual benefits drove this relationship: wolves could scavenge from human campsites, while humans could gain protection and assistance in hunting.

The Process of Domestication:

The process of domestication was gradual and likely involved several stages. Initially, wolves that were less aggressive and more tolerant of humans were drawn to the outskirts of human settlements, attracted by the availability of food scraps. These wolves, with their milder temperaments, were more likely to thrive in proximity to humans and began to exhibit changes in behavior and appearance over generations.

Humans played an active role in this process by selectively encouraging the presence of these more docile wolves. They may have provided food and shelter, further reinforcing the bond between the two species. Over time, this selective pressure resulted in the emergence of dogs, which were distinct from their wild wolf ancestors in both behavior and physical traits.

Early dogs likely had a range of uses in human societies. They served as companions, hunting aids, and sentinels. Their keen senses, particularly their sense of smell and hearing, made them valuable partners in tracking game and detecting danger. As protectors of human campsites, dogs would alert their human counterparts to the presence of intruders or predators, providing an added layer of security.

The Role of Dogs in Early Agricultural Societies:

With the advent of agriculture, the role of dogs became even more significant. As humans transitioned to a more settled lifestyle, dogs adapted to the new environments and needs of farming communities. They helped guard livestock from predators, assisted in herding animals, and continued to provide companionship and protection.

The domestication of dogs had profound implications for the development of human societies. It marked one of the earliest examples of humans successfully domesticating a wild species, setting the stage for the domestication of other animals essential to agricultural life, such as cows, pigs, sheep, and goats.

Dogs' roles diversified as agricultural practices evolved. In addition to their protective and hunting duties, dogs helped manage and control larger herds of animals. This capability was particularly valuable as early farmers began to raise livestock for meat, milk, and other resources.

The Legacy of the First Domesticated Dogs:

The domestication of dogs around 10,000 BCE was a pivotal moment in human history. It showcased the ability of humans to alter their environment and relationships with other species to meet their needs. This successful domestication paved the way for further advancements in agriculture and animal husbandry, contributing to the stability and growth of early farming communities.

Today, dogs remain one of the most versatile and beloved domestic animals, continuing to fulfill roles that have evolved alongside human society. From loyal companions to working animals, their enduring bond with humans is a testament to the deep-rooted relationship that began thousands of years ago.

In summary, the domestication of dogs from wolves around 10,000 BCE was a critical development in the history of agriculture and human civilization. This partnership not only enhanced the survival and efficiency of early farming communities but also laid the foundation for the domestication of other vital species, shaping the trajectory of human progress.

c.8500 BCE: Cows and Pigs

Taming of Cows and Pigs for Meat, Milk, Leather, and Soil Enrichment:

By around 8500 BCE, human societies had made significant strides in domesticating animals, marking another crucial development in the history of agriculture. Among the most important animals to be tamed during this period were cows and pigs. Their domestication provided a variety of resources that fundamentally transformed human life and agricultural practices.

The Domestication of Cows:

Cows, or cattle, were among the first large mammals to be domesticated. This process began in regions such as the Fertile Crescent, where wild ancestors of modern cattle roamed. Early farmers recognized the immense value that these animals could provide, and they began to tame and breed them for specific purposes.

Meat and Milk:
One of the primary reasons for domesticating cows was to secure a reliable source of meat and milk. Cattle meat became a crucial protein source, helping to sustain growing populations. Additionally, cows could be milked, providing a consistent supply of dairy products such as milk, cheese, and yogurt. These dairy products not only offered nutritional benefits but also

allowed for food storage and preservation, contributing to food security.

Leather:

The hides of cattle were used to produce leather, a durable material that was essential for making clothing, footwear, and various tools. Leather's versatility and durability made it a valuable commodity in early agricultural societies.

Soil Enrichment:

Cows also played a vital role in enhancing agricultural productivity through their droppings. Manure from cattle was rich in nutrients and served as an effective natural fertilizer, enriching the soil and improving crop yields. This practice of using animal manure for soil enrichment became a cornerstone of sustainable farming techniques.

The Domestication of Pigs:

Pigs were another species that early farmers began to domesticate around 8500 BCE. The wild ancestors of domestic pigs, known as wild boars, were widespread in regions like the Near East and China. Pigs proved to be particularly adaptable to a variety of environments and provided several key benefits to early agricultural communities.

Meat:

Pigs were primarily domesticated for their meat. Pork became a staple in the diets of many early farming societies, offering a reliable and rich source of protein. The rapid growth and high reproductive rates of pigs made them an efficient and sustainable food source.

Leather:
Like cows, pigs also contributed to the production of leather. Although pigskin was less common than cowhide, it still served as a valuable material for creating clothing and other items.

Soil Enrichment and Waste Management:
Pigs were also instrumental in managing waste and enriching the soil. Pigs are omnivorous and can consume a wide range of food scraps and agricultural by-products, effectively recycling these materials. Their droppings, like those of cows, enriched the soil and helped maintain the fertility of agricultural land.

The Impact of Domesticating Cows and Pigs:

The domestication of cows and pigs had far-reaching impacts on early human societies. These animals provided not only essential resources but also contributed to the development of more complex and stable agricultural systems. By securing reliable sources of meat, milk, and leather, early farmers could focus on other aspects of societal development, such as building permanent settlements and developing trade networks.

The ability to use animal manure as fertilizer improved agricultural productivity, leading to surplus food production. This surplus enabled population growth and the specialization of labor, laying the groundwork for the emergence of advanced civilizations.

Furthermore, the domestication of cows and pigs facilitated the development of social and cultural practices centered around animal husbandry. Rituals, traditions, and economies began to incorporate these animals, reflecting their importance in daily life and in the broader societal context.

In summary, the taming of cows and pigs around 8500 BCE marked a significant milestone in agricultural history. These animals provided crucial resources such as meat, milk, leather, and soil enrichment, transforming early farming practices and contributing to the growth and stability of human societies. Their domestication set the stage for further advancements in agriculture and the continued development of human civilization.

c.7000 BCE: Sheep and Goats

Raising Sheep and Goats for Milk, Food, and Wool Weaving:

By around 7000 BCE, the domestication of sheep and goats had become a pivotal development in early agricultural societies. These animals, well-suited to a variety of environments, provided essential resources that significantly enhanced the livelihoods of early farmers. The domestication of sheep and goats played a crucial role in the evolution of human agriculture, contributing to food security, textile production, and nomadic herding practices.

The Domestication of Sheep:

Sheep were among the earliest animals to be domesticated, primarily for their versatile utility. Their domestication likely began in regions such as the Fertile Crescent and the surrounding highlands, where wild sheep populations were abundant.

Milk:
One of the primary reasons for domesticating sheep was to utilize their milk. Sheep's milk was a valuable source of nutrition and could be used to produce cheese and yogurt, providing a stable and renewable source of food for early farmers. These dairy products could be stored for extended periods, offering sustenance during times when other food sources were scarce.

Food:

Sheep also provided a reliable source of meat. Mutton and lamb became staple foods in many early agricultural societies, offering essential proteins and fats. The relatively small size of sheep compared to larger livestock like cattle made them easier to manage and breed, ensuring a steady supply of meat for growing communities.

Wool Weaving:

Perhaps the most significant contribution of sheep to early human societies was their wool. By around 4000 BCE, people began to realize the potential of wool as a textile fiber. Wool is naturally insulating, durable, and easy to spin into thread, making it an ideal material for clothing and other textiles. The advent of wool weaving marked a significant technological advancement, allowing people to produce warmer, more comfortable, and more durable garments.

The Domestication of Goats:

Goats were domesticated around the same time as sheep and offered many similar benefits. They were well-suited to diverse environments, including rocky and arid regions where other livestock might struggle to survive.

Milk:

Like sheep, goats were primarily valued for their milk. Goat's milk was a crucial dietary component, especially in regions where other dairy sources were less available. It could be consumed fresh or used to make cheese and other dairy products.

Food:
 Goats provided a consistent source of meat. Goat meat, or chevon, was a significant part of the diet in many early farming communities. The high reproductive rates of goats ensured a steady supply of meat, making them a reliable food source.

Mobility and Grazing:
 One of the unique advantages of goats was their adaptability to different environments and their ability to graze on a wide variety of vegetation. Goats could thrive in areas with sparse vegetation, which made them particularly valuable in regions where other livestock might not find sufficient food. This adaptability also supported the nomadic and semi-nomadic lifestyles of many early herding societies, as goats could be easily moved and managed.

The Impact of Raising Sheep and Goats

 The domestication of sheep and goats had a profound impact on early agricultural societies. These animals provided not only essential nutritional resources but also materials that supported the development of new technologies and industries.

Economic and Social Impacts:
 The ability to produce and store surplus milk, meat, and wool led to economic stability and growth. Surpluses could be traded

with neighboring communities, fostering the development of trade networks and economic interdependence. Additionally, the production of wool and dairy products contributed to the specialization of labor, as some community members focused on herding and others on textile production.

Cultural and Technological Advancements:
The domestication of sheep and goats also spurred cultural and technological innovations. The development of wool weaving techniques, for example, revolutionized textile production and clothing. This technological advancement not only improved the quality of life but also supported the growth of trade and the spread of new ideas and practices.

Environmental Adaptations:
Sheep and goats were particularly valuable in regions where other forms of agriculture might be less viable. Their ability to adapt to different environments and their utility in providing multiple resources made them indispensable to early farming communities. The herding of these animals also shaped land use patterns, as pastoralism became an important aspect of agricultural life.

In summary, by around 7000 BCE, the domestication of sheep and goats had become a cornerstone of early agriculture. These animals provided essential resources such as milk, food, and wool, which significantly enhanced the sustainability and prosperity of early farming societies. Their adaptability and versatility made them key contributors to the evolution of human agriculture, supporting the growth of stable and complex communities.

CHAPTER 3: TECHNOLOGICAL INNOVATIONS IN EARLY AGRICULTURE

c.5500 BCE: Irrigation

Development of Irrigation Systems in Mesopotamia:

Around 5500 BCE, one of the most significant advancements in agricultural technology emerged in the region of Mesopotamia: the development of irrigation systems. This innovation transformed agriculture, allowing humans to control water supply for crops and significantly increasing agricultural productivity. The invention and refinement of irrigation systems played a crucial role in the rise of early civilizations in Mesopotamia, a region often referred to as the "Cradle of Civilization."

The Geography and Climate of Mesopotamia:

Mesopotamia, located in the modern-day Middle East, encompassed the area between the Tigris and Euphrates rivers. This region, part of the larger Fertile Crescent, had rich, alluvial soils that were ideal for farming. However, the climate was arid, with hot, dry summers and limited rainfall. The seasonal flooding of the Tigris and Euphrates provided essential water and nutrients to the soil but was unpredictable and often destructive.

To harness the potential of this fertile land, early Mesopotamian farmers needed to find a way to manage and control water resources effectively. This necessity led to the development of irrigation systems, which allowed them to direct water from the rivers to their fields, ensuring a consistent and reliable supply of water for crops.

Early Irrigation Techniques:

The earliest irrigation techniques in Mesopotamia were relatively simple but effective. Farmers dug channels and ditches to divert water from the rivers to their fields. These channels allowed water to flow by gravity, irrigating the land and providing much-needed moisture to crops. This method of surface irrigation, where water spreads across the field, was one of the first steps toward more sophisticated irrigation systems.

Levees and Canals:

As agricultural practices evolved, Mesopotamian farmers began to build more complex irrigation structures, including levees and canals. Levees were embankments constructed along the riverbanks to control flooding and prevent water from inundating fields. By managing the flow of the rivers, farmers could protect their crops from flood damage while also ensuring that water was available for irrigation during dry periods.

Canals were artificial waterways designed to transport water from the rivers to agricultural fields. These canals could be extended over long distances, bringing water to areas that were previously too far from the rivers to be farmed effectively. The construction and maintenance of these canals required considerable effort and cooperation, leading to the development of organized labor and early forms of governance to oversee irrigation projects.

The Impact of Irrigation on Agriculture and Society:

The development of irrigation systems had a profound impact on agriculture in Mesopotamia. By providing a reliable and controlled water supply, irrigation allowed farmers to grow crops in regions that would otherwise be too dry for agriculture. This innovation led to increased agricultural productivity and the ability to cultivate a wider variety of crops.

Irrigation also enabled the cultivation of staple crops such as wheat and barley on a larger scale, supporting the growth of larger, more stable populations. The surplus food produced by irrigated agriculture allowed for the development of trade, the specialization of labor, and the rise of complex societies. As communities grew, so did the need for coordinated management of water resources, leading to the establishment of early forms of government and social organization.

The Legacy of Mesopotamian Irrigation:

The irrigation techniques developed in Mesopotamia laid the foundation for future advancements in agricultural technology. The principles of water management, soil conservation, and crop rotation pioneered by Mesopotamian farmers were passed down through generations and spread to other regions. These innovations contributed to the agricultural success of subsequent civilizations and influenced irrigation practices around the world.

In summary, the development of irrigation systems around 5500 BCE in Mesopotamia was a landmark achievement in the history of agriculture. This innovation allowed early farmers to harness the power of water, transforming arid land into productive farmland and enabling the growth of one of the world's earliest civilizations. The legacy of Mesopotamian irrigation continues to be felt today, as modern irrigation techniques build upon the principles established by these ancient pioneers.

c.5000 BCE: Rice Cultivation

Farming of Rice in Asia and Its Significance as a Staple Food:

Around 5000 BCE, the cultivation of rice began to take root in Asia, marking a pivotal moment in the history of agriculture. This period saw the early farming of rice in regions such as the Yangtze River Valley in China and parts of India. The domestication and cultivation of rice not only revolutionized agricultural practices in Asia but also laid the foundation for the development of complex societies and enduring cultural traditions centered around this staple crop.

The Origins of Rice Cultivation:

Rice, one of the oldest known crops, has a rich history that traces back to the warm, wet regions of Asia. Archaeological evidence suggests that the domestication of rice occurred independently in different regions, with two main subspecies emerging: *Oryza sativa* japonica in the temperate regions of China and *Oryza sativa* indica in the tropical regions of India and Southeast Asia.

The Yangtze River Valley in China is often cited as one of the earliest centers of rice cultivation. Here, ancient farmers began to cultivate wild rice, selectively breeding it for desirable traits such as higher yield and better resistance to environmental conditions. Similarly, in the swampy and riverine areas of India, early agricultural communities started to domesticate rice, adapting it to local climates and soils.

Techniques of Rice Farming:

Early rice farming techniques were labor-intensive but highly effective. Rice requires a significant amount of water to grow, making the floodplains, river valleys, and swampy areas of Asia ideal for its cultivation. The two primary methods of rice farming that emerged were dryland (upland) cultivation and wetland (paddy) cultivation.

Wetland Cultivation:
In wetland or paddy cultivation, rice is grown in flooded fields called paddies. This method involves several steps:
1. Preparation of Seedbeds: Seeds are first sown in specially prepared seedbeds.
2. Transplanting: Once the seedlings are strong enough, they are transplanted to flooded paddies.
3. Flooding: Water is maintained at a consistent level in the paddies to promote growth and control weeds.
4. Harvesting: After several months, the rice plants mature and are harvested.

Dryland Cultivation:
In regions where water was less abundant, rice was grown using dryland methods. This involved planting rice directly in well-prepared soil and relying on rainfall for water. Although less productive than paddy cultivation, dryland farming allowed rice to be grown in more diverse environments.

Significance of Rice as a Staple Food:

Rice quickly became a staple food in many Asian societies due to its high yield and nutritional value. It provided a reliable and abundant source of carbohydrates, essential for sustaining large populations. The cultivation of rice supported the development of dense, stable communities and played a crucial role in the rise of early civilizations in Asia.

Nutritional Benefits:
Rice is rich in carbohydrates and provides essential vitamins and minerals, making it a vital part of the diet. It is also easily digestible, making it suitable for people of all ages.

Economic Impact:
The surplus production of rice allowed for the development of trade networks. Rice became a valuable commodity, traded both within and between early Asian societies. This trade facilitated cultural exchange and the spread of agricultural techniques.

Cultural and Social Influence:
Rice cultivation influenced many aspects of life in ancient Asia, from daily dietary habits to religious and cultural practices. Festivals and rituals celebrating rice harvests became central to many cultures, reflecting the crop's importance.

The Enduring Legacy of Rice Cultivation:

The early farming of rice set the stage for its enduring

significance in Asian culture and cuisine. Today, rice remains a staple food for more than half of the world's population, with Asia producing and consuming the majority of the global supply. The techniques and traditions developed thousands of years ago continue to influence modern rice farming practices.

In summary, the cultivation of rice around 5000 BCE was a transformative development in the history of agriculture. By turning wild rice into a reliable and abundant food source, early Asian farmers laid the groundwork for the growth of complex societies and rich cultural traditions. The legacy of these early agricultural practices continues to shape the world today, underscoring the profound impact of rice cultivation on human history.

CHAPTER 4: AGRICULTURAL TOOLS AND TECHNIQUES

c.200 BCE: Iron Plow

Invention of the Iron Plow in the Han Dynasty:

Around 200 BCE, during the Han Dynasty in China, one of the most revolutionary advancements in agricultural technology was invented: the iron plow. This innovation transformed farming practices, significantly boosting productivity and efficiency, and played a crucial role in supporting the growth and stability of ancient Chinese civilization.

The Agricultural Context of the Han Dynasty:

The Han Dynasty (206 BCE – 220 CE) is often considered a golden age in Chinese history, characterized by significant advancements in science, technology, culture, and economics. Agriculture was the backbone of the Han economy, providing the essential food supply needed to support a large and growing population. Prior to the invention of the iron plow, Chinese farmers used wooden and bronze plows, which were less durable and less effective at breaking up tough soil.

Development and Features of the Iron Plow:

The iron plow was developed as a response to the need for more efficient farming tools that could increase agricultural output. The innovation involved replacing the wooden or bronze components of the plow with iron, a much stronger and more durable material.

Key Features of the Iron Plow:

1. Iron Plowshare:

- The iron plowshare was the key component that cut into the soil. Its sharpness and durability allowed it to break up hard and compacted soil much more effectively than previous plows, making it easier to prepare fields for planting.

2. Adjustable Design:

- Some iron plows featured adjustable components, allowing farmers to control the depth and angle of the plowing. This adaptability was crucial for working different types of soil and terrain, enhancing the versatility of the plow.

3. Moldboard:

- The addition of a moldboard, a curved metal plate, helped to turn over the soil more efficiently, creating furrows and burying weeds. This improved soil aeration and helped to control pests and weeds, promoting healthier crop growth.

Impact on Agriculture and Society:

The invention of the iron plow had profound implications for agriculture in the Han Dynasty and beyond.

Increased Agricultural Productivity:
- The iron plow made it possible to cultivate larger areas of land more quickly and with less physical effort. This increased productivity allowed farmers to produce more food, supporting population growth and urbanization.

Improved Soil Management:
- The iron plow's ability to break up tough soil and turn it over effectively improved soil aeration and fertility. This enhancement in soil quality led to better crop yields and more sustainable farming practices.

Economic and Social Effects:
- The increased agricultural output contributed to economic prosperity and stability. Surplus food production enabled trade and the accumulation of wealth, supporting the development of markets and trade networks. Additionally, the iron plow reduced the labor required for farming, allowing some members of society to pursue other occupations and contribute to cultural and technological advancements.

Spread and Influence:
- The success of the iron plow in China influenced agricultural practices in other regions. As knowledge of this technology spread along trade routes, it was adopted and adapted by various cultures, further enhancing agricultural productivity worldwide.

Legacy of the Iron Plow:

The iron plow represents a significant milestone in the history of agriculture. Its invention during the Han Dynasty not only revolutionized farming practices in ancient China but also set the stage for future innovations in agricultural technology. The principle and benefits of the iron plow continued to be recognized and built upon, contributing to the development of modern farming equipment.

In summary, the invention of the iron plow around 200 BCE during the Han Dynasty was a transformative development in agricultural history. By increasing productivity, improving soil management, and supporting economic and social growth, the iron plow played a critical role in shaping the trajectory of Chinese civilization and influencing agricultural practices around the world.

c.1000 CE: Fieldwork in Western Europe

Crop Rotation Methods in Medieval Europe:

Around 1000 CE, medieval European agriculture underwent a significant transformation with the adoption and refinement of crop rotation methods. This agricultural innovation greatly improved the efficiency and sustainability of farming practices, contributing to population growth, economic development, and the stability of medieval societies.

The Agricultural Context of Medieval Europe:

During the early medieval period, European agriculture was largely characterized by subsistence farming, with small, scattered plots of land cultivated primarily to meet the immediate needs of local communities. Farming techniques were relatively primitive, and the productivity of the land was often limited by poor soil management and inefficient agricultural practices.

The Development of Crop Rotation:

Crop rotation is an agricultural practice that involves growing different types of crops in the same area in sequential seasons. This method helps maintain soil fertility and reduce the buildup of pathogens and pests that often occur when one species is continuously cropped. The three-field system became the predominant crop rotation method in medieval Europe.

The Three-Field System:

- Field One: Food Crops
  - In the first field, farmers would plant food crops such as wheat or rye. These grains were staple foods that provided the primary source of calories and nutrition for the medieval population.

- Field Two: Legumes and Fodder Crops
  - The second field was planted with legumes such as peas, beans, or lentils. These crops not only provided essential protein to the diet but also helped enrich the soil by fixing nitrogen, a crucial nutrient for plant growth. Additionally, fodder crops like oats and barley were grown to feed livestock.

- Field Three: Fallow
  - The third field was left fallow, meaning it was not planted with any crops for a season. This rest period allowed the soil to recover and regain its nutrients, preventing soil depletion and maintaining its fertility.

The rotation typically followed a cycle where each field would be used for a different purpose each year. For example, a field planted with wheat one year would grow legumes the next, and then be left fallow in the third year. This system maximized the productive use of the land while ensuring its long-term sustainability.

Impact on Agriculture and Society:

The introduction of crop rotation methods in medieval Europe had profound and far-reaching effects on agriculture and society.

Increased Agricultural Productivity:
- Crop rotation significantly improved soil fertility and crop yields. By alternating crops and incorporating legumes into the rotation, farmers were able to maintain and enhance the nutrient content of the soil, leading to healthier and more productive crops.

Better Soil Management:
- The three-field system reduced the risk of soil exhaustion and erosion. Leaving a field fallow allowed it to recover, while the use of legumes helped replenish essential nutrients. This careful management of soil resources ensured the long-term viability of agricultural land.

Economic Growth and Stability:
- Higher agricultural productivity supported population growth and urbanization. Surplus food production enabled the growth of markets and trade, contributing to economic development and the accumulation of wealth. The improved efficiency of farming also freed up labor, allowing more people to engage in crafts, trade, and other non-agricultural activities.

Social and Cultural Effects:
- The increased stability and prosperity brought about by improved agricultural practices had significant social and cultural impacts. As communities grew and became more interconnected, cultural exchanges and innovations flourished.

The rise of market towns and trade networks facilitated the spread of new ideas, technologies, and cultural practices.

The Legacy of Medieval Crop Rotation:

The crop rotation methods developed in medieval Europe laid the groundwork for modern sustainable farming practices. The principles of soil fertility management, crop diversity, and efficient land use continue to be fundamental aspects of contemporary agriculture.

In summary, around 1000 CE, the adoption of crop rotation methods, particularly the three-field system, marked a significant advancement in medieval European agriculture. By improving soil fertility and increasing agricultural productivity, crop rotation contributed to the economic growth, social stability, and cultural development of medieval European societies. The legacy of these early agricultural innovations continues to influence farming practices to this day, underscoring the enduring importance of sustainable land management.

CHAPTER 5: GLOBAL EXCHANGE OF CROPS AND ANIMALS

1400s–1500s: Crop Swap

Exchange of Crops and Animals During European Exploration:

The 1400s to 1500s was an era of remarkable global exploration and expansion, driven largely by European powers seeking new trade routes and territories. This period, often referred to as the Age of Exploration, saw not only the discovery of new lands but also the exchange of crops, animals, and agricultural practices between the Old World (Europe, Asia, and Africa) and the New World (the Americas). This exchange, known as the Columbian Exchange, had profound and lasting impacts on agriculture, economies, and diets around the world.

The Age of Exploration:

During the late 15th and early 16th centuries, European explorers, sponsored by monarchs and driven by a desire for wealth, knowledge, and territorial expansion, embarked on voyages across the Atlantic and into the Indian and Pacific Oceans. Notable explorers such as Christopher Columbus, Vasco da Gama, and Ferdinand Magellan led expeditions that connected previously isolated continents.

The Columbian Exchange:

The Columbian Exchange refers to the widespread transfer of plants, animals, culture, human populations, technologies, and ideas between the Americas and the Old World following the voyages of Columbus and other explorers. This exchange dramatically altered agricultural practices and diets on both sides of the Atlantic.

Crops from the Americas to the Old World:
- Maize (Corn):
 - Maize, or corn, was domesticated in Mexico and became a staple food in many parts of the world. It was introduced to Europe, Africa, and Asia, where it quickly became a crucial crop due to its adaptability and high yields.
- Potatoes:
 - Native to the Andes region, potatoes were brought to Europe and thrived in the cool climates of northern Europe. They became a dietary staple and contributed to population growth due to their nutritional value and ability to grow in poor soils.
- Tomatoes:
 - Originating in the Americas, tomatoes were initially met with skepticism in Europe but eventually became integral to cuisines, particularly in Italy and Spain.
- Beans and Squash:
 - Various types of beans and squash, native to the Americas, were introduced to the Old World, adding diversity to European, African, and Asian diets.
- Peppers and Chili Peppers:
 - These spicy fruits, originating in the Americas, spread rapidly across the globe, becoming essential ingredients in cuisines from Asia to Africa to Europe.

Crops from the Old World to the Americas:

- Wheat, Barley, and Rye:
  - These staple grains, essential to European diets, were introduced to the Americas and became fundamental crops in the New World.
- Rice:
  - Rice, originally from Asia, was brought to the Americas, where it was cultivated extensively, particularly in the southern United States.
- Sugarcane:
  - Native to Southeast Asia, sugarcane was introduced to the Caribbean and South America, where the climate was ideal for its cultivation. The sugar industry had significant economic and social impacts, including the development of the transatlantic slave trade.
- Coffee:
  - Originating in Africa, coffee plants were introduced to the Americas and became a major export crop, particularly in regions like Brazil and the Caribbean.
- Citrus Fruits:
  - Oranges, lemons, and limes, native to Asia, were brought to the Americas and thrived in the subtropical climates of the New World.

Animals from the Old World to the Americas:
- Horses:
  - Horses, which had been extinct in the Americas for millennia, were reintroduced by European explorers. They transformed Native American cultures, particularly on the Great Plains, where they became central to hunting, warfare, and

transportation.
- Cattle:
  - Cows, brought by the Spanish and other Europeans, became essential for meat, milk, and labor. Cattle ranching spread throughout the Americas, significantly impacting the economy and landscape.
- Pigs:
  - Pigs, introduced by Europeans, adapted quickly to the American environment, providing a reliable source of meat.
- Sheep and Goats:
  - These animals were also introduced and became important for wool, meat, and milk.

Impact of the Crop Swap:

The Columbian Exchange had profound and lasting impacts on global agriculture, economies, and diets. The introduction of new crops and animals increased agricultural diversity and productivity, supporting population growth and economic development. The exchange also had significant social and cultural implications, as new foods and agricultural practices were integrated into existing traditions and cuisines.

Economic Impact:
- The introduction of high-yield crops like potatoes and maize supported population growth and urbanization in Europe. In the Americas, crops like sugarcane and coffee became major economic drivers, leading to the establishment of plantations and the expansion of the slave trade.

Cultural and Dietary Changes:
- The exchange of crops and animals led to significant changes in diets around the world. Foods that were once regional specialties became global staples, profoundly influencing culinary traditions and food production.

Environmental and Ecological Effects:
- The introduction of new species often disrupted local ecosystems. For example, the proliferation of European livestock in the Americas led to overgrazing and soil degradation in some areas. Conversely, the cultivation of crops like potatoes in Europe transformed agricultural landscapes.

In summary, the period of the 1400s to 1500s marked a transformative era of crop and animal exchange during European exploration. The Columbian Exchange fundamentally altered agriculture, economies, and diets worldwide, with lasting impacts that continue to shape our world today.

CHAPTER 6: THE INDUSTRIAL REVOLUTION AND AGRICULTURE

1794: Cotton Gin

Eli Whitney's Invention and Its Impact on Cotton Production:

In 1794, American inventor Eli Whitney revolutionized the cotton industry with his invention of the cotton gin. This device mechanized the process of separating cotton fibers from their seeds, drastically increasing the efficiency of cotton production. Whitney's cotton gin had profound economic, social, and cultural impacts, particularly in the United States, where it reshaped the agricultural landscape and had far-reaching consequences.

The Problem with Cotton Production:

Before the invention of the cotton gin, cotton production was an incredibly labor-intensive process. Separating the sticky seeds from the cotton fibers by hand was slow and arduous, limiting the amount of cotton that could be processed and making it less profitable as a cash crop. This bottleneck in the production process hindered the growth of the cotton industry, despite the increasing demand for cotton textiles driven by the Industrial Revolution.

Eli Whitney's Invention:

Eli Whitney, a Yale-educated tutor working in the southern United States, was inspired to solve this problem. In 1793, he developed a machine that could efficiently separate cotton fibers from their seeds. By 1794, Whitney had patented his invention, the cotton gin.

How the Cotton Gin Worked:
- The cotton gin consisted of a wooden drum embedded with a series of hooks that pulled the cotton fibers through a mesh. The mesh was too fine for the seeds to pass through, effectively separating them from the fibers.
- A rotating brush then removed the fibers from the hooks, allowing the cleaned cotton to be collected while the seeds were discarded.

The cotton gin could process cotton much faster than manual labor. A single machine could clean as much cotton in one hour as several workers could in a day. This dramatic increase in efficiency revolutionized cotton production.

Impact on Cotton Production:

The cotton gin had an immediate and profound impact on the cotton industry.

Increase in Cotton Production:
- With the efficiency of the cotton gin, cotton became a highly profitable crop. Planters could produce cotton at unprecedented rates, leading to a boom in cotton farming, particularly in the American South.
- The production of cotton skyrocketed. By the mid-19th century, the United States was producing more than 75% of the

world's cotton supply, earning it the nickname "King Cotton."

Economic Growth:
- The explosion in cotton production spurred economic growth in the southern United States. Cotton exports became a major component of the American economy, fueling trade and commerce.
- The increased profitability of cotton also led to investments in infrastructure, such as roads, ports, and railways, to support the expanding cotton trade.

Expansion of Slavery:
- Despite Whitney's intention for the cotton gin to reduce the need for slave labor, it had the opposite effect. The increased demand for cotton led to a corresponding increase in the demand for slave labor to plant, tend, and harvest the cotton.
- The expansion of cotton plantations into the Deep South was accompanied by a significant increase in the slave population. By 1860, there were nearly four million enslaved African Americans in the United States, many of whom worked on cotton plantations.

Social and Cultural Impact:

The cotton gin not only transformed agriculture but also had significant social and cultural implications.

Impact on Society:
- The reliance on slave labor for cotton production entrenched

the institution of slavery in the southern United States, exacerbating sectional tensions between the North and South.
- The profitability of cotton contributed to the economic divide between the industrializing North and the agrarian South, setting the stage for conflicts that would eventually lead to the Civil War.

Cultural Influence:
- Cotton became a symbol of the southern economy and way of life. The wealth generated by cotton production influenced the social hierarchy, with plantation owners becoming powerful economic and political figures.
- The expansion of cotton farming also led to the displacement of Native American communities as land was seized for plantations.

Legacy of the Cotton Gin:

Eli Whitney's cotton gin was a landmark invention that revolutionized agriculture and had lasting impacts on American society. While it greatly enhanced the efficiency and profitability of cotton production, it also had unintended consequences, particularly the entrenchment and expansion of slavery. The cotton gin's legacy is a complex one, reflecting both the technological ingenuity that drives progress and the societal challenges that can arise from such advancements.

In summary, the invention of the cotton gin in 1794 by Eli Whitney was a pivotal moment in agricultural history. It transformed cotton production, fueling economic growth and shaping the social and cultural landscape of the United States. The cotton gin's impact extended far beyond the fields, influencing the course of American history and leaving a lasting legacy on the nation.

1831: Reaping Rewards

Cyrus McCormick's Reaper and Advancements in Harvesting Technology:

In 1831, American inventor Cyrus McCormick revolutionized agriculture with his invention of the mechanical reaper. This device transformed the process of harvesting crops, significantly increasing efficiency and productivity. McCormick's reaper was a critical advancement in agricultural technology, leading to profound changes in farming practices and contributing to the agricultural boom of the 19th century.

The Challenges of Harvesting Before the Reaper:

Before the invention of the mechanical reaper, harvesting grain was an arduous and labor-intensive process. Farmers relied on hand tools such as sickles and scythes to cut down crops, a method that was slow, physically demanding, and inefficient. The time-consuming nature of manual harvesting limited the amount of land that could be cultivated and constrained the potential for increased agricultural output.

Cyrus McCormick's Invention:

Cyrus McCormick, born into a family of inventors in Virginia, recognized the need for a more efficient method of harvesting grain. Building on earlier attempts by his father, Robert McCormick, Cyrus developed a successful mechanical reaper in

1831. His reaper featured several key components that made it effective and practical for widespread use.

Key Features of the McCormick Reaper:
- Cutting Mechanism:
  - The reaper used a reciprocating blade to cut the stalks of grain. This blade moved back and forth rapidly, slicing through the crop as the machine moved forward.
- Reel:
  - A rotating reel helped to guide the standing grain towards the cutting mechanism, ensuring an even and efficient cut.
- Platform:
  - The cut grain fell onto a platform, where it could be collected and bundled by workers following the reaper. This arrangement streamlined the process of gathering and transporting the harvested grain.

The mechanical reaper could perform the work of several laborers, greatly increasing the speed and efficiency of the harvest. McCormick's reaper allowed farmers to harvest more acres of grain in less time, significantly boosting productivity.

Impact on Agriculture and Society:

The introduction of McCormick's reaper had far-reaching impacts on agriculture and society, reshaping the agricultural landscape and contributing to economic growth.

Increased Agricultural Productivity:

- The reaper enabled farmers to cultivate and harvest much larger areas of land. This increase in efficiency allowed for the expansion of farms and the cultivation of more crops, leading to higher yields and greater overall production.
- The ability to harvest grain quickly and efficiently reduced the risk of crop loss due to weather or other factors, providing greater security and stability for farmers.

Economic Growth:
- The increased productivity brought about by the reaper contributed to economic growth and prosperity. Surplus grain could be sold in markets, fueling trade and commerce.
- The success of the reaper led to the establishment of manufacturing plants to produce the machines, creating jobs and stimulating industrial growth. McCormick's company, the McCormick Harvesting Machine Company, became a leading manufacturer of agricultural equipment.

Social and Cultural Effects:
- The reaper reduced the need for manual labor in the fields, freeing up workers to pursue other occupations and contributing to the growth of urban centers. This shift facilitated the transition from an agrarian economy to a more industrialized society.
- The increased efficiency of farming practices also supported population growth, as higher agricultural productivity could sustain larger communities.

Impact on the American Frontier:

- The reaper played a significant role in the expansion of agriculture into the American frontier. As settlers moved westward, the reaper allowed them to cultivate and harvest vast tracts of land more effectively, contributing to the settlement and development of the western United States.

Legacy of the McCormick Reaper:

Cyrus McCormick's invention of the mechanical reaper was a pivotal moment in the history of agriculture. The reaper not only revolutionized the process of harvesting grain but also set the stage for further advancements in agricultural technology. The principles of mechanization and efficiency embodied in the reaper influenced the development of subsequent farming equipment, leading to continuous improvements in agricultural practices.

The legacy of the McCormick reaper is evident in the modern agricultural machinery that continues to drive productivity and innovation in farming. McCormick's contributions to agriculture earned him a place as one of the great inventors of the 19th century, and his reaper remains a symbol of the transformative power of technology in shaping human history.

In summary, the invention of the mechanical reaper by Cyrus McCormick in 1831 was a transformative advancement in harvesting technology. The reaper significantly increased the efficiency and productivity of grain harvesting, contributing to economic growth, societal change, and the expansion of agriculture. McCormick's reaper stands as a testament to the enduring impact of technological innovation on the agricultural industry and society at large.

1837: Steel Plow

John Deere's Invention of the Steel Plow for the American Prairie:

In 1837, blacksmith John Deere revolutionized agriculture with the invention of the steel plow, a tool that transformed farming practices on the American prairie. Deere's innovation addressed the unique challenges of prairie farming and significantly boosted agricultural productivity, paving the way for the expansion and development of American agriculture.

The Challenges of Prairie Farming:

Before the invention of the steel plow, farmers on the American prairie faced significant difficulties in cultivating the tough, dense soil known as prairie sod. Traditional iron and wooden plows used in the Eastern United States were ill-suited for these conditions, as the sticky soil would cling to the plow blades, making it difficult to break the ground and slowing down the work considerably. This inefficiency hindered the agricultural potential of the vast and fertile prairie lands.

John Deere's Innovation:

John Deere, a blacksmith who had moved from Vermont to Illinois, recognized the need for a more effective plow to handle the challenging prairie soil. Drawing on his skills and experience, Deere designed a new type of plow made from polished steel, which was much more effective at cutting through and turning over the tough sod.

Key Features of the Steel Plow:
- Polished Steel Blade:
  - The blade of Deere's plow was made from high-quality, polished steel. This material had a smooth surface that allowed the soil to slide off easily, preventing the plow from becoming clogged and enabling it to cut through the soil more efficiently.
- Curved Moldboard:
  - Deere's design included a curved moldboard that helped to lift and turn over the soil. This feature was crucial for creating clean furrows and preparing the land for planting.

Deere constructed his first steel plow in 1837, and its effectiveness was immediately apparent. The plow worked so well that it became known as the "self-polishing plow" because it cleaned itself as it moved through the soil, maintaining its efficiency throughout the plowing process.

Impact on Agriculture and Society:

The introduction of John Deere's steel plow had a profound impact on agriculture, particularly in the Midwest, where the fertile prairies held great potential for farming.

Increased Agricultural Productivity:
- The steel plow made it much easier and faster to break up the dense prairie sod, allowing farmers to cultivate larger areas of land with less effort. This increased productivity led to higher

crop yields and greater overall agricultural output.
- The efficiency of the steel plow also reduced the labor required for plowing, freeing up time and resources for other farming activities.

Expansion of Farming:
- The ability to efficiently plow the prairie soil encouraged the westward expansion of American agriculture. Farmers were able to settle and cultivate new lands, contributing to the growth of rural communities and the development of the agricultural economy.
- The expansion of farming into the prairies played a crucial role in the settlement and development of the American Midwest, transforming the region into a major agricultural hub.

Economic Growth:
- The increased agricultural productivity supported economic growth and prosperity. Surplus crops could be sold in markets, fueling trade and commerce and contributing to the overall economic development of the United States.
- Deere's invention also had a significant impact on the manufacturing industry. As demand for the steel plow grew, Deere established a company to produce and distribute the plows, leading to the growth of the agricultural machinery industry. The company he founded, Deere & Company, became one of the largest and most successful agricultural equipment manufacturers in the world.

Legacy of the Steel Plow:

John Deere's invention of the steel plow in 1837 was a transformative moment in agricultural history. The steel plow not only revolutionized farming on the American prairie but also set the stage for future advancements in agricultural technology. The principles of efficiency, durability, and innovation embodied in Deere's plow continue to influence the design and development of modern farming equipment.

The legacy of the steel plow is evident in the continued success and influence of Deere & Company, which remains a leader in the agricultural machinery industry. John Deere's contributions to agriculture earned him a lasting place in the history of American innovation, and his steel plow remains a symbol of the transformative power of technological advancement in agriculture.

In summary, the invention of the steel plow by John Deere in 1837 was a pivotal development in agricultural technology. The steel plow addressed the unique challenges of prairie farming, significantly increasing productivity and facilitating the expansion of American agriculture. Deere's innovation transformed farming practices, contributed to economic growth, and left an enduring legacy on the agricultural industry and society.

CHAPTER 7: SCIENTIFIC ADVANCES IN AGRICULTURE

1866: Selective Breeding

Gregor Mendel's Experiments and Their Influence on Crop Breeding:

In 1866, the field of genetics and the practice of selective breeding were revolutionized by the pioneering work of Gregor Mendel, an Augustinian monk whose meticulous experiments with pea plants laid the foundation for modern genetics. Mendel's discoveries about inheritance patterns provided critical insights that influenced crop breeding, leading to more efficient and targeted agricultural practices.

Gregor Mendel's Experiments:

Gregor Mendel conducted his groundbreaking experiments in the garden of his monastery in Brno, Austria (now the Czech Republic). Over several years, Mendel cultivated and cross-bred thousands of pea plants (*Pisum sativum*), meticulously recording the outcomes of his experiments. His work focused on understanding how traits were passed from one generation to the next.

Key Aspects of Mendel's Experiments:

1.Choice of Pea Plants:
   - Mendel chose pea plants because they had several distinct, easily observable traits (such as flower color, seed shape, and pod color), and they could be easily manipulated for controlled

breeding experiments.

2. Controlled Cross-Pollination:
   - By carefully cross-pollinating plants with different traits, Mendel was able to track how these traits were inherited in successive generations. He would manually transfer pollen from one plant to the stigma of another, ensuring precise control over the mating process.

3. Observation of Trait Inheritance:
   - Mendel observed that certain traits appeared in predictable ratios in the offspring. For example, when he crossed plants with purple flowers and white flowers, the first generation (F1) exhibited only purple flowers. However, in the second generation (F2), both purple and white flowers appeared in a 3:1 ratio.

4. Formulation of Mendel's Laws:
   - Based on his observations, Mendel formulated two fundamental laws of inheritance:
   - Law of Segregation: Each organism carries two alleles for each trait, which segregate during the formation of gametes. Offspring thus inherit one allele from each parent.
   - Law of Independent Assortment:** Alleles for different traits are distributed to gametes independently, resulting in genetic variation.

Influence on Crop Breeding:

Although Mendel's work was not widely recognized during his

lifetime, it was rediscovered in the early 20th century and quickly became the cornerstone of modern genetics. His principles of inheritance had profound implications for crop breeding, leading to more systematic and scientifically informed practices.

Selective Breeding:
- Mendel's discoveries allowed breeders to understand how traits were inherited and to predict the outcomes of specific crosses. This knowledge enabled the development of selective breeding programs aimed at enhancing desirable traits such as yield, disease resistance, and environmental adaptability.

Hybridization:
- The understanding of genetic inheritance paved the way for hybridization, where two genetically different parent plants are crossbred to produce hybrid offspring with superior traits. Hybrid crops often exhibit "hybrid vigor" or heterosis, resulting in greater productivity and resilience.

Genetic Diversity:
- Mendel's work highlighted the importance of maintaining genetic diversity within crop populations. Breeders could now intentionally introduce new traits and genetic variation to improve crops and adapt them to changing environmental conditions and agricultural demands.

Marker-Assisted Selection:
- With the advancement of genetic technology, Mendel's principles have been applied in marker-assisted selection (MAS).

This technique uses molecular markers linked to desirable traits to identify and select plants with those traits more efficiently, speeding up the breeding process.

Impact on Global Agriculture:
- The application of Mendelian genetics in crop breeding has led to the development of high-yielding, disease-resistant varieties of staple crops such as wheat, rice, maize, and soybeans. These advancements have contributed significantly to global food security and agricultural sustainability.

Legacy of Mendel's Work:

Gregor Mendel's meticulous experiments and the principles of inheritance he discovered laid the groundwork for modern genetics and revolutionized crop breeding. His insights enabled the systematic improvement of crops, enhancing their productivity, resilience, and nutritional value. The continued application of Mendelian genetics in agricultural science ensures that Mendel's legacy endures, driving innovation and progress in crop breeding and agriculture.

In summary, the year 1866 marks a pivotal moment in the history of agriculture and genetics with Gregor Mendel's experiments on pea plants. His discoveries about inheritance patterns provided the scientific foundation for selective breeding, transforming crop breeding practices and significantly impacting global agriculture. Mendel's work continues to influence the development of new and improved crop varieties, underscoring the enduring importance of his contributions to science and agriculture.

1885: Combine Harvester

Hugh Victor McKay's Invention and Its Impact on Farming Efficiency:

In 1885, Australian inventor Hugh Victor McKay revolutionized agriculture with his invention of the combine harvester. This machine combined three essential harvesting operations—reaping, threshing, and winnowing—into a single process, dramatically increasing farming efficiency and transforming agricultural practices worldwide.

The Challenges of Traditional Harvesting:

Before the invention of the combine harvester, harvesting grain was a labor-intensive and time-consuming process. Farmers had to cut the crop (reaping), separate the grain from the stalks (threshing), and then clean the grain by removing the chaff (winnowing). Each step required significant manual labor and time, limiting the amount of land that could be cultivated and harvested efficiently.

Hugh Victor McKay's Innovation:

Hugh Victor McKay, an innovative farmer and inventor from Victoria, Australia, recognized the need for a more efficient harvesting method. In 1885, at the age of 22, he developed the first commercially successful combine harvester, which he called the Sunshine Harvester.

Key Features of McKay's Combine Harvester:**

1. Integration of Functions:
 - The combine harvester integrated reaping, threshing, and winnowing into a single machine. This integration allowed the grain to be cut, separated, and cleaned in one continuous operation as the machine moved through the field.

2. Mechanization:
 - The machine was powered initially by horses and later by steam engines and internal combustion engines. This mechanization significantly reduced the need for manual labor, allowing a single operator to harvest large areas of crops efficiently.

3. Efficiency:
 - The combine harvester greatly increased the speed and efficiency of the harvesting process. What previously took days or weeks to accomplish by hand could now be done in a fraction of the time, freeing up labor and resources for other agricultural tasks.

Impact on Farming Efficiency:

The introduction of the combine harvester had profound and far-reaching effects on farming efficiency and agricultural productivity.

Increased Agricultural Productivity:

- The combine harvester allowed farmers to harvest their crops much more quickly and efficiently. This increased productivity meant that larger areas of land could be cultivated, leading to higher overall crop yields.
- The ability to harvest crops rapidly reduced the risk of crop loss due to adverse weather conditions, ensuring that more of the planted grain could be successfully gathered.

Labor Savings:
- The mechanization of the harvesting process significantly reduced the need for manual labor. This labor-saving aspect allowed farmers to reallocate their workforce to other important tasks, such as planting and maintenance, further enhancing overall farm productivity.
- The reduction in labor requirements also helped address labor shortages, particularly during peak harvest seasons.

Economic Impact:
- The efficiency gains from using combine harvesters contributed to the economic growth of agricultural regions. Farmers were able to produce more grain with fewer resources, increasing their profitability and stimulating local economies.
- The success of McKay's combine harvester also spurred the growth of the agricultural machinery industry, leading to advancements in other types of farming equipment and technologies.

Global Adoption:
- The benefits of the combine harvester quickly became

apparent, leading to its adoption worldwide. Agricultural regions in North America, Europe, and beyond began using combine harvesters, transforming global farming practices and increasing food production on a large scale.

Legacy of the Combine Harvester:

Hugh Victor McKay's invention of the combine harvester in 1885 was a landmark achievement in agricultural history. The combine harvester revolutionized the way grain was harvested, greatly increasing efficiency and productivity. McKay's invention set the stage for continuous improvements in agricultural machinery and farming practices, contributing to the modernization of agriculture.

The impact of the combine harvester continues to be felt today. Modern combine harvesters, now equipped with advanced technologies such as GPS guidance systems and computerized controls, are even more efficient and capable than their predecessors. These machines remain a critical component of large-scale farming operations, underscoring the enduring legacy of McKay's pioneering work.

In summary, the invention of the combine harvester by Hugh Victor McKay in 1885 significantly improved farming efficiency by integrating reaping, threshing, and winnowing into a single mechanized process. This innovation transformed agricultural practices, increased productivity, and had a lasting impact on the agricultural industry and global food production.

1890: Tractor

John Froelich's Rudimentary Tractor and Its Significance:

In 1890, American inventor John Froelich made a groundbreaking advancement in agricultural technology with the creation of the first successful gasoline-powered tractor. This innovation marked the beginning of a new era in farming, transforming agricultural practices and significantly enhancing efficiency and productivity.

The Challenges of Traditional Farming:

Before the advent of tractors, farming operations were largely dependent on human and animal labor. Horses and oxen were the primary sources of power for plowing, planting, and other fieldwork. While effective, these methods were labor-intensive, time-consuming, and limited by the physical capabilities of the animals. Additionally, the maintenance and feeding of draft animals were significant expenses for farmers.

John Froelich's Innovation:

John Froelich, a farmer and inventor from Iowa, recognized the limitations of animal-powered farming and sought to develop a more efficient solution. In 1890, he successfully built the first gasoline-powered tractor, which he initially referred to as a "gasoline traction engine."

Key Features of Froelich's Tractor:

1. Gasoline Engine:

 - Froelich's tractor was powered by a one-cylinder gasoline engine. This engine provided a reliable and consistent source of power, overcoming the limitations of steam engines, which were cumbersome and required significant preparation time to build up steam.

2. Mobility and Versatility:

 - The tractor was mounted on a chassis with wheels, allowing it to move easily across fields. This mobility made it versatile for various agricultural tasks, such as plowing, threshing, and hauling.

3. Ease of Use:

 - Froelich's design included features that made the tractor user-friendly and practical for farmers. It was equipped with a forward and reverse transmission, giving it greater maneuverability.

Froelich tested his invention by using it to power a threshing machine during the harvest season. The success of this trial demonstrated the potential of gasoline-powered tractors to revolutionize farming practices.

Significance of the Tractor:

The introduction of John Froelich's gasoline-powered tractor had profound and lasting impacts on agriculture, paving the way for the widespread adoption of mechanized farming.

Increased Efficiency and Productivity:
- The tractor significantly increased the efficiency of farming operations. It could perform tasks such as plowing and threshing much faster and more effectively than horses or oxen, allowing farmers to cultivate larger areas of land and improve overall productivity.

Reduced Labor and Costs:
- The use of tractors reduced the reliance on human and animal labor, decreasing the physical demands on farmworkers and lowering labor costs. Additionally, farmers no longer needed to maintain and feed large teams of draft animals, reducing operational expenses.

Enhanced Agricultural Practices:
- Tractors enabled the adoption of more advanced agricultural techniques and machinery. With a reliable source of power, farmers could use a wider variety of implements, such as seed drills, harrows, and reapers, further increasing the efficiency and effectiveness of their work.

Economic Impact:
- The introduction of tractors contributed to economic growth in rural areas. Increased agricultural productivity led to higher yields and greater profitability for farmers, stimulating local economies. The manufacturing and maintenance of tractors also created new industries and job opportunities.

Environmental Benefits:
- Tractors allowed for more precise and timely fieldwork, improving soil management and crop yields. The ability to plow fields more effectively reduced soil erosion and promoted better soil health, contributing to sustainable farming practices.

Legacy of the Tractor:

John Froelich's invention of the gasoline-powered tractor in 1890 was a pivotal moment in agricultural history. The tractor revolutionized farming practices, making agriculture more efficient, productive, and sustainable. Froelich's innovation laid the groundwork for the development of modern tractors, which continue to be indispensable tools in agriculture worldwide.

Today, tractors are equipped with advanced technologies such as GPS navigation, computerized controls, and environmental sensors, making them even more efficient and effective. The legacy of John Froelich's pioneering work endures in the continued innovation and advancement of agricultural machinery, underscoring the transformative impact of the tractor on farming and society.

In summary, the invention of the gasoline-powered tractor by John Froelich in 1890 marked a significant advancement in agricultural technology. The tractor's ability to increase efficiency, reduce labor, and enhance agricultural practices transformed farming, contributing to economic growth and sustainable agriculture. Froelich's innovation remains a cornerstone of modern farming, highlighting the enduring importance of mechanization in agriculture.

CHAPTER 8: MODERN AGRICULTURE

1940: Green Revolution

Introduction of Modern Farming Practices and Technologies:

The Green Revolution, which began in the 1940s, marked a major turning point in agricultural history. This period was characterized by the introduction of new farming practices and technologies that dramatically increased agricultural productivity and food production, especially in developing countries. The Green Revolution aimed to combat hunger and improve food security by modernizing agriculture.

Key Components of the Green Revolution:

High-Yielding Varieties (HYVs):
- One of the cornerstones of the Green Revolution was the development and dissemination of high-yielding varieties (HYVs) of staple crops such as wheat, rice, and maize. These new varieties were bred to produce more grain per plant, significantly increasing crop yields.
- The introduction of HYVs was spearheaded by agricultural scientists like Norman Borlaug, whose work on developing disease-resistant, high-yielding wheat varieties earned him the Nobel Peace Prize in 1970.

Chemical Fertilizers and Pesticides:
- The use of chemical fertilizers and pesticides became widespread during the Green Revolution. Fertilizers provided essential nutrients to crops, enhancing their growth and

productivity, while pesticides helped control pests and diseases that could devastate crops.
- These inputs were essential for maximizing the potential of HYVs, ensuring that plants had the necessary nutrients and protection to thrive.

Irrigation Infrastructure:
- Improved irrigation infrastructure was another critical component of the Green Revolution. The construction of dams, canals, and other irrigation systems allowed farmers to control water supply more effectively, ensuring that crops received adequate moisture throughout the growing season.
- Reliable irrigation reduced the dependence on unpredictable rainfall and allowed for multiple cropping cycles per year, further boosting productivity.

Mechanization:
- The adoption of modern agricultural machinery, such as tractors, combine harvesters, and pumps, played a significant role in the Green Revolution. Mechanization reduced the labor required for planting, tending, and harvesting crops, increasing efficiency and allowing farmers to cultivate larger areas of land.
- These technologies also helped reduce the time needed for farming operations, enabling quicker turnaround between planting and harvesting cycles.

Impact of the Green Revolution:

Increased Agricultural Productivity:

- The Green Revolution led to dramatic increases in crop yields and overall agricultural productivity. Countries that adopted these new practices and technologies saw significant improvements in food production, which helped reduce hunger and improve food security.
- For example, in India and Mexico, wheat and rice production more than doubled within a few decades, transforming these countries from food-deficient to self-sufficient and even surplus-producing nations.

Economic Growth:
- The surge in agricultural productivity contributed to economic growth, particularly in rural areas. Increased food production supported the development of agribusinesses, created jobs, and generated income for millions of farmers.
- The Green Revolution also stimulated rural economies by increasing demand for agricultural inputs such as fertilizers, pesticides, and machinery, as well as for services related to storage, transportation, and marketing of agricultural products.

Social and Environmental Effects:
- While the Green Revolution brought many benefits, it also had social and environmental consequences. The focus on high-input farming practices led to the intensification of agriculture, which in some cases resulted in soil degradation, depletion of water resources, and increased pollution from chemical fertilizers and pesticides.
- Additionally, the Green Revolution tended to benefit wealthier farmers who could afford the new technologies and inputs,

leading to increased economic disparities in some regions. Smallholder and subsistence farmers often struggled to compete, and traditional farming practices and crop varieties were sometimes displaced.

Legacy of the Green Revolution:

The Green Revolution had a profound and lasting impact on global agriculture. It demonstrated the potential of scientific and technological advancements to transform food production and combat hunger. The principles and practices developed during this period continue to influence modern agriculture, with ongoing efforts to address the challenges and limitations identified in the original Green Revolution.

In summary, the Green Revolution of the 1940s introduced modern farming practices and technologies that revolutionized agriculture. The adoption of high-yielding crop varieties, chemical fertilizers and pesticides, improved irrigation, and mechanization led to significant increases in agricultural productivity and food security, particularly in developing countries. While the Green Revolution brought substantial benefits, it also highlighted the need for sustainable and equitable approaches to agricultural development.

1990: Genetically Modified (GM) Crops

Development and Controversy Surrounding GM Crops:

In the 1990s, the advent of genetically modified (GM) crops marked a significant milestone in agricultural biotechnology. GM crops were developed using advanced genetic engineering techniques to introduce new traits into plants, such as pest resistance, herbicide tolerance, and improved nutritional content. While these innovations promised to enhance agricultural productivity and sustainability, they also sparked significant controversy and debate over their safety, environmental impact, and ethical implications.

Development of GM Crops:

Genetic Engineering Techniques:
- Genetic modification involves altering the DNA of a plant by inserting genes from other organisms. These genes can come from a variety of sources, including bacteria, viruses, or other plants, to confer specific desirable traits.
- Techniques such as recombinant DNA technology and CRISPR-Cas9 allowed scientists to precisely modify plant genomes, leading to the development of crops with enhanced characteristics.

Pioneering GM Crops:
- The first commercially grown GM crop was the Flavr Savr tomato, approved for sale in the United States in 1994. It was

engineered for a longer shelf life by delaying the ripening process.
- Other early GM crops included Bt corn, which was modified to express a toxin derived from the bacterium Bacillus thuringiensis to protect against insect pests, and Roundup Ready soybeans, which were engineered to tolerate glyphosate herbicide, allowing farmers to control weeds without damaging the crop.

Advantages of GM Crops:
- Pest Resistance: GM crops such as Bt corn reduce the need for chemical insecticides, lowering production costs and minimizing environmental impact.
- Herbicide Tolerance: Crops like Roundup Ready soybeans allow for more effective weed control, reducing competition for resources and increasing yields.
- Improved Nutritional Content: Biofortified crops, such as Golden Rice enriched with vitamin A, aim to address nutritional deficiencies in developing countries.
- Drought and Salinity Tolerance: Researchers have developed crops that can withstand extreme environmental conditions, helping to ensure food security in the face of climate change.

Controversy and Debate:

Safety Concerns:
- Critics of GM crops raised concerns about their safety for human consumption. Potential issues included the risk of allergenicity, gene transfer to non-target species, and unintended effects on human health.

- Regulatory agencies such as the U.S. Food and Drug Administration (FDA), the European Food Safety Authority (EFSA), and others conducted extensive assessments to evaluate the safety of GM crops before approving them for market.

Environmental Impact:
- The environmental effects of GM crops were also a major point of contention. Concerns included the potential for gene flow to wild relatives, leading to the development of "superweeds" resistant to herbicides, and the impact on non-target organisms, such as beneficial insects and soil microbes.
- Supporters argued that GM crops could reduce the need for chemical inputs, decrease soil erosion through reduced tillage, and contribute to more sustainable agricultural practices.

Ethical and Socioeconomic Issues:
- Ethical debates centered on the morality of altering the genetic makeup of living organisms and the potential long-term consequences.
- Socioeconomic concerns included the control of seed patents by large biotechnology companies, which critics argued could disadvantage small farmers and reduce biodiversity by promoting monocultures.

Public Perception and Acceptance:
- Public perception of GM crops varied widely, with acceptance generally higher in the United States and lower in Europe and parts of Asia. Misinformation and lack of understanding about genetic engineering contributed to public skepticism and

resistance.

- Advocacy groups, both for and against GM crops, played significant roles in shaping public opinion and influencing regulatory policies.

Legacy and Future of GM Crops:

Ongoing Research and Development:
- Despite the controversies, research and development of GM crops continued to advance. New techniques such as CRISPR-Cas9 have enabled even more precise genetic modifications, potentially addressing some of the earlier concerns.
- The focus has expanded to include traits such as enhanced nutritional content, improved shelf life, and resistance to environmental stresses.

Regulatory and Policy Developments:
- Regulatory frameworks for GM crops have evolved, with increased emphasis on transparency, safety assessments, and environmental monitoring.
- International trade policies and agreements have also addressed the movement and labeling of GM products, reflecting the global nature of agricultural markets.

Integration with Sustainable Practices:
- There is a growing recognition of the need to integrate GM crops with broader sustainable agricultural practices. This includes promoting crop diversity, conserving natural resources, and ensuring equitable access to technology for all farmers.

In summary, the development of genetically modified (GM) crops in the 1990s represented a major technological advancement in agriculture. While GM crops offered significant benefits in terms of pest resistance, herbicide tolerance, and improved nutrition, they also generated considerable controversy and debate over safety, environmental impact, and ethical issues. The legacy of GM crops continues to influence modern agriculture, with ongoing research, regulatory developments, and efforts to integrate biotechnology with sustainable farming practices.

CONCLUSION

Recap of the Evolution of Agriculture:

The history of agriculture is a testament to human ingenuity and adaptability. From the dawn of farming over 10,000 years ago in the Fertile Crescent to the modern technological advancements that shape today's agricultural practices, humanity has continually innovated to meet the growing demands for food and resources.

Early Beginnings:
- The transition from hunting and gathering to settled farming around 11,000 BCE marked the beginning of agriculture. Early farmers in regions such as the Fertile Crescent domesticated plants and animals, creating the foundation for stable communities and the development of civilizations.

Technological Innovations:
- The domestication of key animals like dogs, cows, pigs, sheep, and goats between 10,000 and 7,000 BCE provided essential resources such as meat, milk, wool, and labor. The development of irrigation systems around 5500 BCE in Mesopotamia and the cultivation of rice in Asia around 5000 BCE further advanced agricultural productivity.

Agricultural Tools and Techniques:
- Innovations such as the iron plow during the Han Dynasty (c. 200 BCE) and crop rotation methods in medieval Europe (c.

1000 CE) improved soil management and increased yields. The exchange of crops and animals during European exploration in the 1400s and 1500s enriched global agriculture.

Industrial Revolution and Beyond:
- The 19th century saw the invention of key agricultural machinery, including Eli Whitney's cotton gin (1794), Cyrus McCormick's reaper (1831), and John Deere's steel plow (1837), which revolutionized farming efficiency and productivity. The development of the combine harvester by Hugh Victor McKay in 1885 and John Froelich's gasoline-powered tractor in 1890 further transformed agricultural practices.

Scientific Advances:
- The 20th century brought significant scientific advancements, such as Gregor Mendel's principles of genetics in 1866, which laid the groundwork for selective breeding. The Green Revolution of the 1940s introduced modern farming practices and technologies, dramatically increasing food production. The development of genetically modified (GM) crops in the 1990s introduced new possibilities and controversies in agriculture.

Reflection on the Future of Farming and Food Production:

As we look to the future, agriculture faces both significant challenges and exciting opportunities. The global population is projected to reach nearly 10 billion by 2050, increasing the demand for food and placing pressure on agricultural systems. Climate change, resource scarcity, and environmental degradation further complicate the task of ensuring food security for all.

Sustainable Practices:
- The future of farming will require a focus on sustainability. This includes adopting practices that conserve natural resources, protect biodiversity, and reduce the environmental impact of agriculture. Techniques such as precision farming, organic agriculture, and agroecology offer pathways to more sustainable food production.

Technological Innovations:
- Advances in technology will continue to play a crucial role in the evolution of agriculture. Developments in biotechnology, such as CRISPR and other gene-editing tools, hold the promise of creating crops that are more resilient to climate change and pests, and more nutritious. Digital technologies, including artificial intelligence, drones, and the Internet of Things (IoT), can optimize farming practices, improve efficiency, and reduce waste.

Climate Resilience:
- Building resilience to climate change is essential for the future of agriculture. This includes developing drought-resistant and flood-tolerant crop varieties, improving soil health, and adopting water-efficient irrigation practices. Farmers will need to adapt to changing weather patterns and increased incidences of extreme weather events.

Food Systems and Policy:
- The future of food production also depends on creating equitable and inclusive food systems. Policies that support smallholder farmers, promote fair trade, and ensure access to nutritious food for all are critical. Addressing food waste and improving supply chain efficiency are also important components of a sustainable food system.

Global Collaboration:
- Addressing the challenges of future food production requires global collaboration and knowledge sharing. International organizations, governments, researchers, and farmers must work together to develop and implement innovative solutions. Investing in agricultural research and education is vital to equip future generations with the skills and knowledge needed to sustainably feed the world.

In conclusion, The history of Agriculture is a story of continuous innovation and adaptation. From the earliest days of farming to the cutting-edge technologies of today, agriculture has always been at the heart of human progress. As we face the future, embracing sustainable practices, harnessing technological advancements, and fostering global collaboration will be key to ensuring that agriculture continues to thrive and provide for the growing needs of humanity.

GLOSSARY OF TERMS

A

Agriculture: The science, art, and practice of cultivating soil, growing crops, and raising animals for food, fiber, and other products used to sustain and enhance human life.

Agroecology: An ecological approach to agriculture that views agricultural areas as ecosystems and is concerned with the ecological impact of agricultural practices.

B

Biodiversity: The variety of plant and animal life in a particular habitat, essential for the stability and resilience of ecosystems.

Biofortification: The process of increasing the nutritional value of food crops through biological means, such as plant breeding or genetic modification.

C

Combine Harvester: A machine that combines the tasks of reaping, threshing, and winnowing crops into a single process, significantly improving harvesting efficiency.

Crop Rotation: The practice of growing different types of crops in the same area in sequential seasons to improve soil health and reduce pest and disease problems.

CRISPR: A gene-editing technology that allows for precise modifications to an organism's DNA, widely used in genetic engineering.

D

Domestication: The process of adapting wild plants and animals for human use, including agriculture, by selecting for desired traits over many generations.

F
Fallow: Agricultural land left unseeded for a period to restore its fertility as part of crop rotation or to avoid surplus production.

G
Genetic Engineering: The direct manipulation of an organism's genes using biotechnology to create new traits or enhance existing ones.
Genetically Modified (GM) Crops: Crops that have been altered using genetic engineering techniques to introduce desirable traits such as pest resistance, herbicide tolerance, or improved nutritional content.

H
High-Yielding Varieties (HYVs): Crop varieties that have been selectively bred or genetically modified to produce higher yields under certain conditions, often used in the Green Revolution.

I
Irrigation: The artificial application of water to land to assist in the growth of crops, an essential practice for agriculture in arid and semi-arid regions.

M
Mechanization: The use of machinery to perform agricultural tasks, which increases efficiency and productivity while reducing the need for manual labor.

P
Precision Farming: An agricultural management concept that uses GPS, sensors, and other technologies to optimize field-level management regarding crop farming.

R

Reaper: A machine or person that harvests grain or other crops, notably advanced by Cyrus McCormick in the 19th century.

Roundup Ready Crops: Genetically modified crops that are resistant to glyphosate, the active ingredient in the herbicide Roundup, allowing for easier weed control.

S

Selective Breeding: The process of breeding plants and animals for particular genetic traits, based on the principles of genetics discovered by Gregor Mendel.

Sustainability: Farming practices that meet current food needs without compromising the ability of future generations to meet their own needs, focusing on environmental health, economic profitability, and social equity.

T

Tractor: A powerful motor vehicle with large rear wheels, used mainly on farms for hauling equipment and trailers, first successfully developed by John Froelich in 1890.

W

Winnowing: The process of separating grain from chaff, traditionally done by throwing the mixture into the air so that the wind blows away the lighter chaff while the heavier grain falls back down.

TIMELINE OF KEY EVENTS

c.11,000–9000 BCE: First Farmers
- Development of farming in Syria and Iran after the last Ice Age.
- Growth of wheat and barley in the Fertile Crescent.

c.10,000 BCE: A Farmer's Best Friend
- Domestication of dogs from wolves.

c.8500 BCE: Cows and Pigs
- Taming of cows and pigs for meat, milk, leather, and soil enrichment.

c.7000 BCE: Sheep and Goats
- Raising sheep and goats for milk, food, and wool weaving.

c.5500 BCE: Irrigation
- Development of irrigation systems in Mesopotamia.

c.5000 BCE: Rice Cultivation
- Farming of rice in Asia and its significance as a staple food.

c.200 BCE: Iron Plow
- Invention of the iron plow in the Han Dynasty.

c.1000 CE: Fieldwork in Western Europe
- Crop rotation methods in medieval Europe.

1400s–1500s: Crop Swap
- Exchange of crops and animals during European exploration.

1794: Cotton Gin
- Eli Whitney's invention and its impact on cotton production.

1831: Reaping Rewards
- Cyrus McCormick's reaper and advancements in harvesting technology.

1837: Steel Plow
- John Deere's invention of the steel plow for the American prairie.

1866: Selective Breeding
- Gregor Mendel's experiments and their influence on crop breeding.

1885: Combine Harvester
- Hugh Victor McKay's invention and its impact on farming efficiency.

1890: Tractor
- John Froelich's rudimentary tractor and its significance.

1940: Green Revolution
- Introduction of modern farming practices and technologies.

1990: Genetically Modified (GM) Crops
- Development and controversy surrounding GM crops.

REFERENCES AND FURTHER READING

Books

1. "Guns, Germs, and Steel: The Fates of Human Societies" by Jared Diamond
   - This book explores the factors that influenced the development of human societies, including the role of agriculture.

2. "An Edible History of Humanity"** by Tom Standage
   - Standage provides a detailed history of agriculture and its impact on human civilization.

3. "The Omnivore's Dilemma: A Natural History of Four Meals" by Michael Pollan
   - Pollan investigates the food chain that sustains us, examining the origins and implications of what we eat.

4. "Agriculture: A Very Short Introduction" by Paul Brassley and Richard Soffe
   - This concise book offers an overview of the history and current state of agriculture.

5. "The Green Revolution: An International Bibliography" by Michael P. Collinson
   - A comprehensive bibliography that covers the literature on the Green Revolution.

Articles and Papers

1. "The Origins of Agriculture" by David R. Harris and Gordon C. Hillman
   - This paper explores the early development of agriculture and domestication.

2. "Green Revolution: Impacts, limits, and the path ahead"** by Prabhu L. Pingali
   - An analysis of the successes and limitations of the Green Revolution, along with future directions.

3. "Genetically Modified Crops: Mechanisms and Environmental Risks"** by Guy R. Knudsen and John P. Marshall
   - A scientific review of the development, benefits, and risks associated with GM crops.

4. "The Role of Mechanization in Agriculture" by the Food and Agriculture Organization (FAO)
   - A comprehensive report on the impact of mechanization on agricultural productivity and sustainability.

Online Resources

1. The Food and Agriculture Organization of the United Nations (FAO)
   - Website: www.fao.org
   - The FAO provides extensive resources and publications on global agricultural practices and innovations.

2. The International Rice Research Institute (IRRI)
   - Website: www.irri.org
   - IRRI focuses on the development and dissemination of high-yielding rice varieties and sustainable farming practices.

3. The Norman Borlaug Institute for International Agriculture
   - Website: borlaug.tamu.edu
   - Named after the "father of the Green Revolution," this institute promotes agricultural research and education.

4. US Department of Agriculture (USDA)
 - Website: www.usda.gov
 - The USDA offers a wealth of information on agricultural policies, research, and innovations in the United States.

Journals

1. Agricultural Systems
 - This journal publishes articles on the design and implementation of sustainable agricultural systems.

2. Field Crops Research
 - A journal focusing on the improvement of field crops through modern agricultural practices and technologies.

3. Journal of Agricultural and Food Chemistry
 - This journal covers the latest research on the chemistry and biochemistry of agriculture and food production.

4. Nature Biotechnology
 - A leading journal that publishes research on biotechnology, including advancements in genetic modification and crop improvement.